CryptoNovus
CRYPTIONARY

Cryptocurrency and Blockchain Lingo for Newbies

I0032346

AUTHOR

Christopher Brown

Published by

FaithFinance using
Blurb Book Wright Software

Published in the United States of America

ISBN: **978-1-7345500-2-3**

1. Personal Finance
2. Finance
3. Cryptocurrency Technology

DEDICATION

This book is dedicated to my all my fellow Noviciusians and
Dominusians in The World of CryptoNovus
where …..

" Newbies become Masters "
" Novicius – fit – Dominus "

SPECIAL ACKNOWLEDGMENTS

I would like to thank my wife Allison for the patience and time I
needed to write this book.

Table of Contents

Forward 5

The Lingo of Digital Assets 7

Acronyms 13

Cryptionary 23

Digital Money, A New Frontier 51

      History of Digital Cash 52

      Digital Money & Bitcoin Today 58

      Digital Money's Evolution 60

Author's Bio 64

FORWARD

If you are new to blockchain and cryptocurrency you have a variety of books and information sources to help you with understanding the lingo of this new industry. I wanted to take the time and compile as many of the relevant acronyms and definitions I could think which have helped me as a "Newbie" in this industry.

I also wanted to give a little more and explain why this is important and well as give a little insight as to how long this "crypto thang" has been going on and where it could lead us globally. This is certainly not an academic manuscript and is an easy read. However, for the investor it will take your level of comprehension up a notch or two which will in turn equipped you to make wiser investment decisions.

May God bless you on your stewardship journey and investment decisions.

This book is divided into three parts

Acronyms: an easy-to-follow reference to meanings of the main abbreviations used in crypto conversation and the world of crypto trading.

Cryptionary (Dictionary): explanations for the key words, terms, and slang used in the realm of cryptocurrency buying and selling.

Explanation: taking a closer look at some of the topics of money which help make the fundamentals of cryptocurrency technology possible.

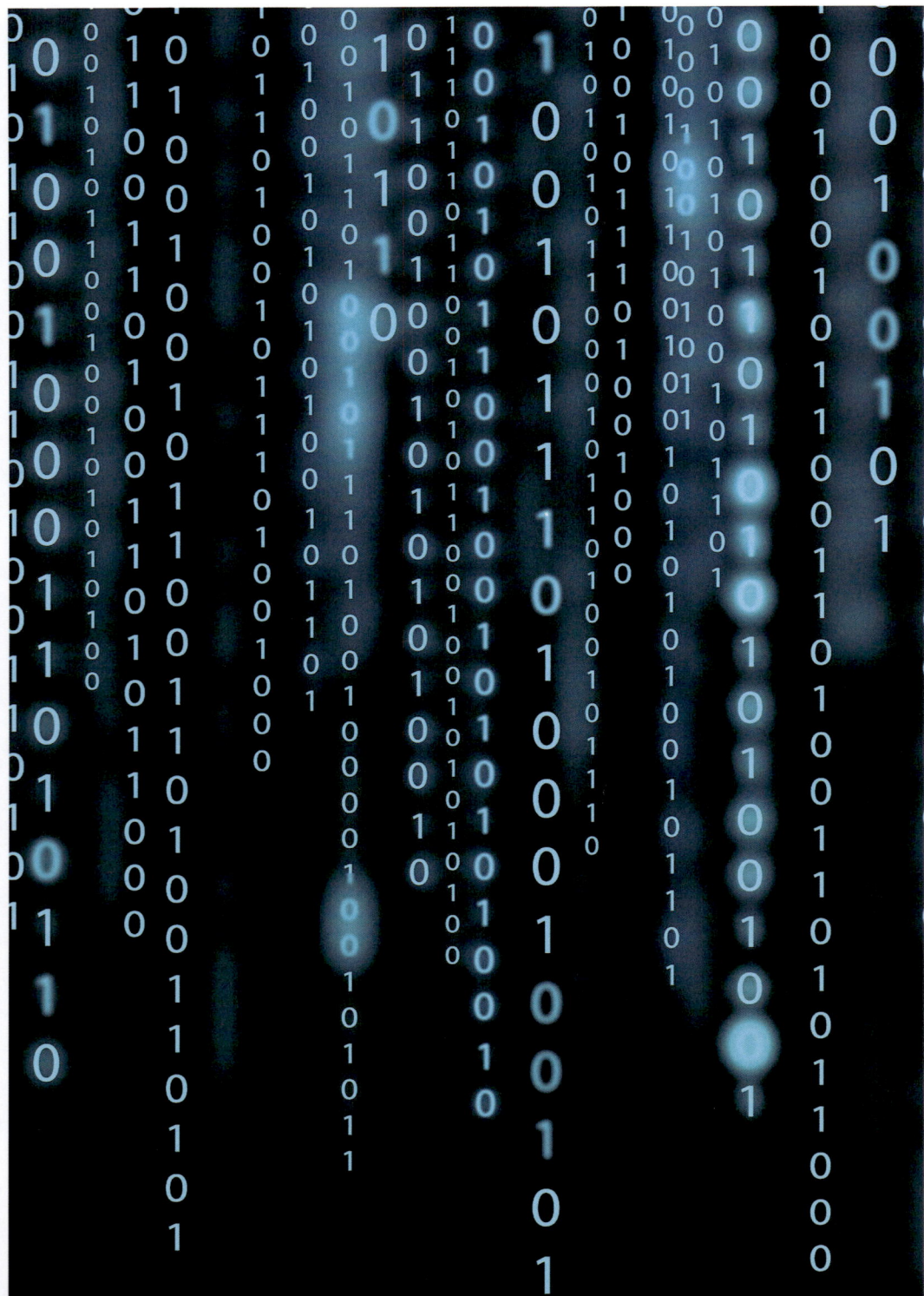

The Lingo
of Digital Assets

As new technologies are introduced to the world, they inevitably create a series of new terms and definitions which requires learning a new "language." An example of this, is how the introduction of the internet has expanded our vocabulary over the decades. I remember when I first heard the phrase, "going online" and thinking, "on what?" Since then, we have been introduced to phrases like "smart phones", "smart houses", "mobile app" and "Dapps" which have become part of our everyday conversations.

Technology now brings us to the world of Digital Assets and its language of **cryptography.** The word crypto is derived from the Greek "**kryptos**", which translates as "**hidden, concealed, or secret**". Cryptography means the art or writing or solving codes and is a common term used by those intellectuals accustomed to the realm of computer science. Simply put, cryptography is a method of writing code for storing and transmitting data in a form that can be read and processed only by those for whom it is intended. This modern technology has been used for decades in government, the military, and commercial communication, and data centers.

Cryptography has been instrumental in developing what are popularly referred to as **"cryptocurrencies."** This very new asset class has been evolving since 1991 when Stuart Haber and W. Scott Stornetta envisioned what many people have come to know as blockchain. technology. Their first work involved working on a cryptographically secured chain of blocks whereby no one could tamper with timestamps of documents. The market of Cryptocurrencies represents a significant leap forward for digital technology and for money itself. In this space, cryptography has been predominately used to create a secure and private pipeline between sender and receiver, such as HTTPS websites. There are many great minds, such as Italian computer scientist, Silvio Micali (aka The Grandfather of Cryptography), who understand the power of this technology and continue discovering new ways to use it beyond the world of finance.

Satoshi Nakamoto, the pseudonym to describe the person or group, is accredited as the brains behind first cryptocurrency to capture public attention. The blockchain technology behind Bitcoin is the first application of the digital ledger technology. Nakamoto's 2008 white paper, *"Bitcoin: A Peer-to-Peer Electronic Cash System"* outlined the architecture for a **peer-to-peer (P2P) network** where participants would engage in computer **"mining"** operations, essentially requiring all the computers or **nodes** in the network to compete to solve a computationally intensive mathematical problem. The first to successfully solve the problem would be awarded in bitcoin. The creation, support and maintenance of the network would establish direct interaction and verification, or **proof-of-work**, among the participants. This is revolutionary because this process called **"trust minimizing transactions"** essentially avoids the need for transactions to pass through third parties, such as financial institutions.

"Trust minimizing" in this instance reveals to the novel framework that supports bitcoin and other digital currencies: the **blockchain**. A blockchain consists of a series of connected ledgers, upon which are recorded the specific details of every transaction executed on the system. One of the most notable features of the blockchain is its ability to permanently prevent duplication or alteration of any kind. Transaction details recorded on a block prevent a bitcoin, or other digital asset, from being copied or counterfeited. This is just one of several security layers provided by a blockchain system. The key here is that the system prevents the use of the same digital token for more

8

than one transaction simultaneously. The details added to the blocks are verified by all the computers in the network.

The blockchain functions essentially as a **decentralized ledger,** a key construct in a digital currency network such as Bitcoin. One of the primary factors behind the creation of Bitcoin, and other digital token networks, was a desire to develop a currency that could not be controlled by a central bank or government. As we know, the value of the U.S. dollar is manipulated by the Federal Reserve all the time to adjust for inflation.

Bitcoin, on the other hand, is an independent network without any centralized control. The network operates through a decentralized, global system made up of thousands of computers known as **nodes** which as of January of 2021 there were nearly 83,000 (this number fluctuates) which communicate, transmit, and verify bitcoin transactions and data to each other.

Public key cryptography is another important method that is used to ensure secure communication and use of a digital token such as bitcoin. In short, public keys are shared with others and allow encrypted messages or instructions to pass between parties. A public key is shared with others with whom you wish to communicate. Conversely, a private key is used to privately compute the public key and is used to decrypt or verify a message or instruction, such as spending bitcoin. **This private code is only known by the owner and should never be shared**.

Digital signatures and multiple signatures (Multi-Sig) use public key cryptography to verify messages, in a way that is far superior to handwritten signatures. In the digital environment, your signature proves that it is your signature, but also proves that you signed a *specific document!* Multiple signatures are commonly used when more than one person is responsible for creating and approving a transaction, sometimes referred to as a **maker/checker** procedure. This is a preferred institutional regimen to ensure that a transaction is properly authorized by requiring more than one signature before executing a transaction, mitigating risks associated with a single person or device being able to send a transaction.

The bitcoin token was designed to be "digital money", serving as a global medium of exchange and function as a **"store of value."** Since Bitcoin's inception other cryptocurrencies have been developed to do the same, such as Litecoin, Monero and Zcash. While others, such as Ethereum and Enjin, have been developed as network platforms. How digitally-produced currencies or tokens gain, hold, and lose value remains an open question which is being defined as the digital asset market matures and adoption occurs.

A major factor supporting bitcoin's role as a possible store of value is the fixed limit that is placed on the number of bitcoins that will ever be mined. The network will only mine 21 million bitcoins between now

and the year 2140, with the output managed by the Bitcoin network year by year. The "miner's reward" for this output is reduced by half every four years which also has an impact on Bitcoin's unit value.

Another support for bitcoin's use as a potential longer-term store of value relates to the cost of the energy required to create a bitcoin. The large and powerful computers on the Bitcoin network are specifically built to solve a complex mathematical computation and win bitcoin. The average **electricity costs** to run the network's computers are often cited as one measure of the "value" of a bitcoin, just as the costs of mining gold, silver or diamonds are reflected in their values respectively.

In this maverick new frontier of digital assets there are many new words, term and phrases to learn which describe this new technology, how to use it and just as important how to invest in it. As this emerging asset class matures, even more lingo and newly-minted words will also emerge to describe the products, processes, and applications within the digital coin and token realm. The next few pages will begin to expose you the acronyms and definitions of Lingo of Digital Assets.

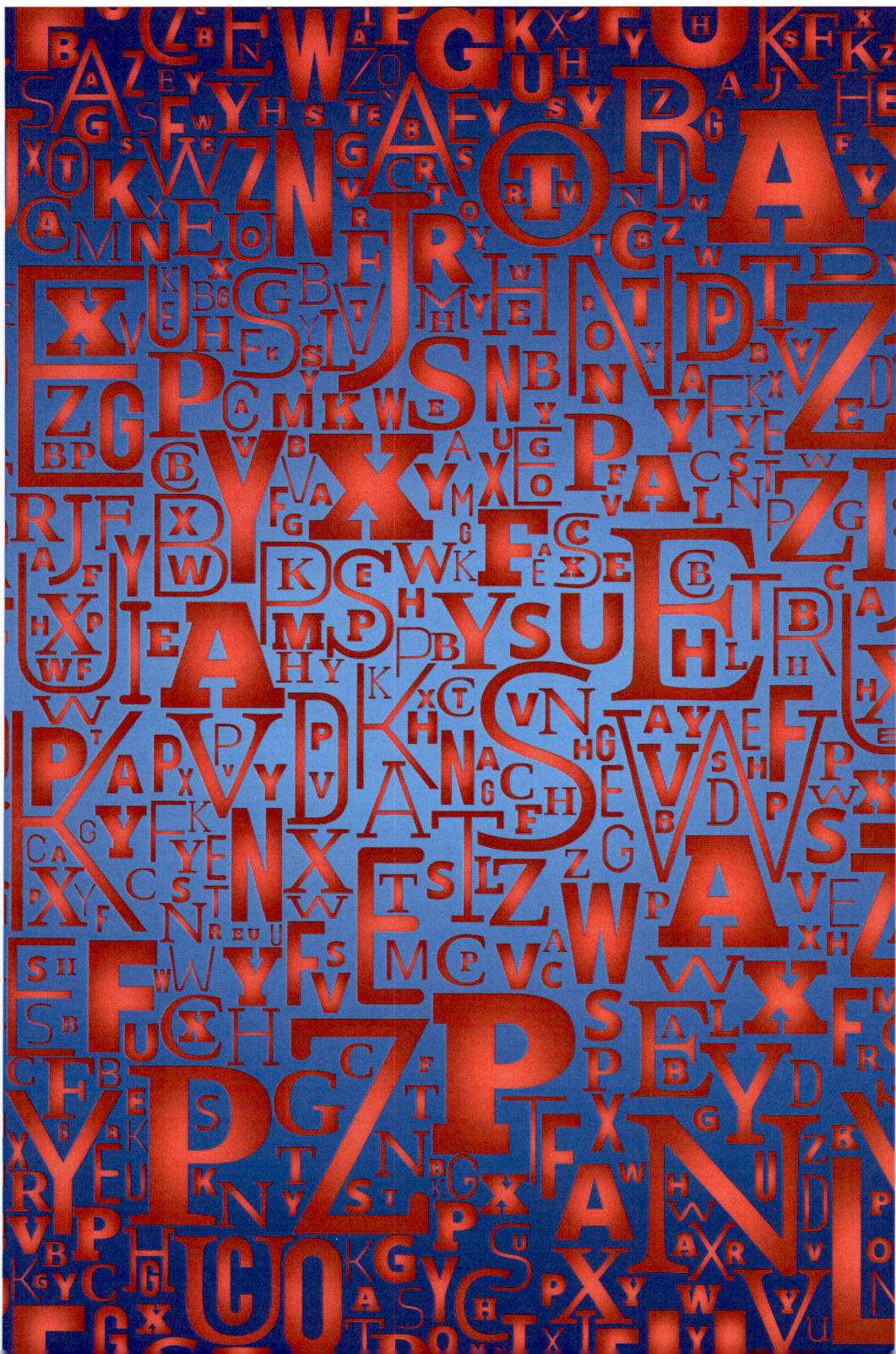

Acronyms

A

AB	Accounting Blockchain
AAL	Account Abstract Layer
ACL	Ada Crypto Library
AEAD	Authenticated Encryption with Associated Data
AES	Advanced Encryption Standard
AOTB	Art on the Blockchain
AWC	Asian Wealth Community
AWC	Atomic Wallet Coin
AC	Alternate Currency
ADC	Association of Digital Currency
AML	Anti-Money Laundering
API	Application Programming Interface
ASIC	Application Specific Integrated Circuit
ATH	All Time High
ATL	All Time Low

B

BCC	Blockchain Competence Center
BCN	Blockchain Consensus Network
BCSO	Bitcoin Cash Standard Organization
BFT	Byzantine Fault Tolerance
BGN	Blockchain Global News
BIA	Blockchain Interoperability Alliance
BIC	Blockchain Investors Consortium
BIOT	Blockchain Internet of Things
BPA	Branch Prediction Analysis
BPFI	Banking and Payments Federation of Ireland

BWW	Blockchain World Wire
BIP	Bitcoin Improvement Proposal
BPI	Bitcoin Price Index
BTC	Bitcoin
BTD	Buy The Dip
BU	Bitcoin Unlimited

C

CAB	Crypto Agent Bot
CBP	Certified Bitcoin Professional
CCL	Cardano Computer Layer
CE	Crypto Engine
CAB	Crypto Agent Bot
CBP	Certified Bitcoin Professional
CENS	Crypto Exchange Nova Service
CER	Crypto Exchange Ranks
CESC	Crypto Economics Security Conference
CENS	Crypto Exchange Nova Service
CFTC	U.S. – Commodity Futures Trading Commission
CGCX	Calvin Global Crypto Exchange
CGF	Crypto Growth Factor
CIAT	Cryptographic Implementation Analysis Toolkit
CIK	Crypto Ignition Key
CIS	Cryptocurrency Investor Summit
CML	Crypto Millions Lotto
CO	Crypto Officer
CS	Crypto Suite
CSP	Cryptographic Service Provider
CSRNG	Cryptographically Secure Random Number Gen.
CTA	Cyber Threat Alliance
CTB	Curve To Bitcoin
CTL	Crypto Trade Line

CTP	Crypto Trader's Pro
CVA	Crypto Valley Association
CVLL	Crypto Variable Logic Label
CW	Crypto Word
cBTC	centiBitcoin

D

DAC	Decentralized Autonomous Company
DAICO	Decentralized Autonomous ICO
DAO	Decentralized Autonomous Organizations
DAPPs	Decentralized Applications
DCA	Dollar Cost Averaging
DDV	Distributed Data Vending
DDoS	Districubuted Denial of Service
DEX	Decentralized Exchange
DOS	Denial Of Service
DGE	Digital Gold Exchange
DTX	Digital Ticks Exchange
DYOR	Do Your Own Research
dBTC	deciBitcoin
daBtc	decaBitcoin

E

ECB	European Crypto Bank
ECC	Elliptic Curve Crypto
EDF	Ecosystem Development Fund
EEA	Enterprise Etherum Alliance
EEAS	Enterprise Ethereum Architecture Stack
EIP	Esports Interactive Platform
EIP	Ethereum Improvement Proposal
ELI5	Explain Like I am 5

EMA	Exponential Moving Average
ERC	Etherum Request for Comments
ERC-20	Ethereum Request for Comments / Number
ESP	Embedded Service Processor
ETF	Exchange Traded Fund
ETH	Ethereum
ETT	Eternal Trust Token
EVT	Ethfinex Voting Token
EVm	Ethereum Virtual Machine
EW	Elliot Wave

F

FA	Fundamental Analysis
FATCA	Foreign Account Tax Compliance Act
FBA	Federated Byzantine Agreement
FCT	Free Crypto Token
Flash P&D	Flash Pump and Dump
FOMO	Fear Of Missing Out
FreeCO	Free Coin Offering
FTEC	First Trading Ecosystem Cryptocurrency
FUD	Fear Uncertainty Doubt

G

GBE	Gibraltar Blockchain Exchange
GBS	Global Blockchain Summit
GCH	Global Crypto Hub
GDCM	Galaxy Digital Capital Management

H

HDAC	Hyundai Digital Asset Company

HF	Hard Fork
HODL	Hold Your Investment
HT	Huobi Tokens
H&S	Head and Shoulders

I

IA	Issued Assets
IBO	Initial Bunty Offering
IC0	Initial Coin Offering
IDAEC	International Digital Assets Exchange Council
IE0	Initial Exchange Offering
ILP	Initial Loan Procurement
IMT	Initial Money Token
IMO	In My Opinion
IOC	Immediate or Cancel
ISO	Initial Scam Offering
ITF	Initial Token Offerings

J

JCE	Java Crypto Extensions
JVCEA	Japanese Virtual Currency Exchange Assoc.
JVCEA	Japan Virtual Currency Exchange Agency

K

KBA	Korea Blockchain Association
KNC	Kyber network Crystals
kBTC	Kilo Bitcoin (10000)

L

L2L	Lan to Lan
LN	Lighting Network

M

MA	Moving Averages
MACD	Moving Average Convergence / Divergence Indicator
MASF	Miner Activated Soft Fork
MBTC	Megabitcoin (1,000,000 btc)
mBTC	Millibitcoin (.001 BTC)
MCAP	Marketcapilization
MCT	My Crypto Trade
MEW	My Ethereum Wallet
MMC	Monthly Maintenance Charge
MPCA	My Paying Crypto Ads
MPP	Multi-Party Payment Protocol
MPCA	My Paying Crypto Ads
MSE	Maltese Stock Exchange

N

NABC	North American Bitcoin Conference
NADDIC	Never A Dull Day In Crypto
NFT	Non-Fungible Tokens
NGC	NEO Global Capital
NTUC	National Trading Union Congress

O

OSC	Open Science Chain

P

PBFT	Practical Byzantine Fault Tolerance
PHI	Platform for Hybrid Investments
PKA	Public Key Accelerator
PKC	Public Key Crypto
PKI	Public Key Infrastructure

PoA	Proof of Authority
PoI	Proof of Importance
PoS	Proof of Stake
PoW	Proof of Work
PSBT	Partially Signed Bitcoin Transactions
P&D	Pump and Dump
P2P	Peer to Peer

Q

QA	Qualitative Analysis
QBT	Qiwi Blockchain Technologies
QSDI	Quantum Secure Digital Identity

R

RBF	Replace By Fee
RingCT	Ring Confidential Transactions
RNG	Random Number Generation
ROI	Return on Investment
RoRo	Risk on, Risk off
RPOW	Reusable Proofs of Work
RSI	Relative Strength Index

S

SAFT	Simple Agreement for Future Tokens
SATS	Satoshi's
SEC	Securities and Exchange Commission
SegWit	Segregated Witness
SF	Soft Fork
SHA	Secure Hash Algorithm

SoV	Store of Value
SPV	Simple Payment Verification
STO	Security Token Offering
SCJS	Society for Crypto-Judaic Studies
SJCL	Stanford Javascript Crypto Library
SPMT	Secure PIVX Masternode Tool
STO	Securties Token Offering

T

TA	Technical Analysis
TCF	Token Classification Framework
TCR	Token-Curated Registry
TGE	Token Generation Event
TOR	Terms of Reference
TPS	Transactions Per Second
TT	Trial Token
TTD	Time To Dump
TWAP	Time Weighted Average Price

U

U2F	Universal 2-Factor
UAHF	User Activated Hard Fork
UASF	User Activated Soft Fork
uBTC	microBitcoin (.000001)
UTXO	Unspent Transaction Output Per Second

V

VaR	Value at Risk
VCA	Virtual Commodity Association

W

WCC	World Crypto Coin
WCEF	World Crypto Economic Forum

```
ne = "
ataLoader

equestRespon

equestResponse

equestResponse Execu

questResponse GetTask

cionSchedule();

cionScheduleItem();

ecutionSchedule
```

Cryptionary

A

Address – A string of code that refers to the wallet of an individual who is sending or receiving cryptocurrency. The address is an individual's digital avatar in a cryptocurrency network. An address is a "payment instruction" for a digital asset. When receiving payment, a payee communicates an address to the payor, and the payor sends funds to that address. A set of addresses used together comprises a wallet.

Altcoin – A general term used to refer to all "alternative" cryptocurrencies besides Bitcoin. This coins and tokens were developed after the success of Bitcoin. Many altcoins are legitimate projects with real business use cases. However, there are many which are illegitimate and have little practical use or trust in the broader public. They are sometimes hard to transfer and may also be outright scams.

Arbitrage – A trading strategy where a trader takes advantage of different prices for the same product on different exchanges. While arbitrage is most common for stocks and currencies, it can also be applied to cryptocurrencies.

ATH – All-time high. This term refers to the highest-ever price of an asset like a cryptocurrency. Bitcoin has hit this number repeatedly over the past two years.

B

Bagholder – A term used for a cryptocurrency investor who buys and holds coins in large quantity with the optimism that the price will increase. However, the investor is left with a worthless digital asset after the successful completion of a pump-and-dump scheme. They are the victims of these scams.

Bear Market – A market where the long-term prices of an asset such as a cryptocurrency are decreasing. Investors during a bear market are more cautious and pessimistic than usual.

Bearish – A negative attitude about an asset that the price will soon drop or continue to drop.

Bitcoin / bitcoin – Bitcoin (with a capital B) is the original form of cryptocurrency launched in 2009. It established the world's first decentralized digital asset. Bitcoin uses blockchain technology to create a digital asset managed across a wide network of computers. The popularity of the term "Bitcoin" has made it shorthand for all forms of cryptocurrency. The virtual coins generated by the Bitcoin network are called bitcoin (with a lowercase b). A "bitcoin" can be used as a medium of exchange or be a "store of value." While digital assets are speculative and present risks, the longevity and saturation of the Bitcoin and Ethereum networks, and their coins, have made them leading candidates for product support such as custody and execution services.

Bitcoin Whale – Is an early Bitcoin investor who bought a large quantity of Bitcoin when the value was relatively low. As a result, this investor can influence market pricing.

Block – A block is a set of updates to the blockchain ledger. Using Bitcoin as an example, a block is basically a virtual container of bitcoin transactions. A block can hold a limited amount of data allowing for a certain number of transactions and the corresponding data to be stored in each block. A bitcoin node receives these blocks, validates all transactions in them and then applies the updates to the global ledger.

A bitcoin miner is tasked to validate all transaction in the block then solve a complicated mathematical equation that cryptographically ties this block to previous blocks. Once broadcast to other nodes and miners, this block is added to the string of blocks that make up the chain. The whole blockchain is a publicly viewable record that keeps track of every transaction that has ever occurred within that digital asset. Simply put, the collection of data (transactions) which are grouped together for verification and eventually added to the blockchain.

Block Height – This is a number that specifies how many blocks have been globally produced at the present time. The very first block created in a blockchain (known as the *genesis block*) has a height of zero because it is the first block in the chain. The fifth block to be added will have a height of four because four blocks came before it. As of October 2018, the Bitcoin block height is almost 550,000.

Block Reward – This term refers to the reward that a miner receives for processing a part of the blockchain that verifies a transaction. Block rewards are the way in which miners receive cryptocurrency. The first miner to solve the proof-of-work puzzle in a block receives a block reward of new coins as compensation for the miner's expenditure in solving the puzzle. In the case of bitcoin, the reward given is cut in half every four years to control the distribution of coins released. Other altcoins do something similar.

Blockchain – A system of ledgers protected by cryptography that computers in a network can access. Blockchain is the underlying technology that Bitcoin and most other digital assets use to record and validate transactions. It is a linked list of transaction updates to a virtual digital public ledger. A blockchain consists of a group of transactions in *blocks*. These blocks are cryptographically connected to one another as they are mined, creating a long *chain*. The nature of the cryptographic tie from one block to previous blocks means that previous blocks cannot be altered by anyone.

BTC – Is the original shorthand for *bitcoin*. This designation is often used on digital asset exchanges to denominate a bitcoin's current value. However, there has been an increase in the use of XBT as an alternate designation. The reason for this is that the International Organization for Standardization (ISO), which keeps a listing of all currencies, uses *X* to symbolize a currency that is not attached to a specific country (which is the case for all digital assets, because they are decentralized).

Bull Market – A market where the value of an asset has increased and is projected to increase in the future. This market causes investors to become more optimistic and go long more than usual.

Bullish – An attitude about any kind of asset that the asset will increase in value over the short-term and the long-term.

C

Circulating Supply – A best guess of the total number of token or coins in circulation. This is usually used when computing the coin's "market cap". I say best guess because it's almost impossible to calculate accurately.

Coin – A representation of an asset that resides on its own blockchain (like bitcoin).

Coinbase – One of the largest and most trustworthy cryptocurrency exchanges available today. A sign of legitimacy in the world of cryptocurrency is listing a currency on this exchange.

Cold Storage – The way by which individuals take their cryptocurrency out of online systems so it can be kept away from possible hacking. It is a storage mechanism where private keys used to

sign withdrawal transactions are kept in secure locations that are not connected to the internet.

Cryptocurrency – An anonymous, decentralized form of currency based on lines of code that make up a blockchain. Digital currency or money, that utilizes encryption and cryptography to control the generation of new units of currency as well as secure and verify transactions of that currency. Cryptocurrencies, also known as *digital assets* and *digital currencies*, are issued and transferred electronically.

Cryptocurrency Exchange – Is a digital trading marketplace where traders can buy and sell bitcoin and altcoins (i.e. ETH, LTC, ADA) using different fiat currencies. It is an online trading platform that acts as an intermediary between buyers and sellers of the cryptocurrency. Cryptocurrency exchanges are similar to stock exchanges where buyers and sellers are matched.

Cryptojacking – A process where a computer is taken over and surreptitiously used to mine cryptocurrency. Cryptojacking often uses only a small amount of processing power from several hundred or thousand computers for illicit mining.

Custody – A service in which a financial institution or other entity holds property on behalf of a customer.

D

DAO – Stands for Dentralized Autonomous Organization. It was a venture capital fund built on the Ethereum network that was hacked in 2016, losing almost a third of its funds. The DAO is often referred to in the news when highlighint the risks of cryptocurrency.

Dapp – Stands for Decentrazlied Application. This is an application that uses blockchain technology.

Decentralized – A system in which there is no central authority or entity that holds control, but rather the system's resources and processes are distributed among many entities.

Digital Asset – A description for this emerging asset class. Several other terms, such as *cryptocurrencies*, *crypto assets*, *virtual currencies*, and *crypto tokens*, are also used in this evolving market.

Digital Signature – Is a mechanism that uses public-key cryptography to create un-forgeable proof that a transaction is authorized by the owner of the coins. The most common algorithm used by digital assets is the Elliptic Curve Digital Signature Algorithm (ECDSA), though there are many such algorithms, including Schnorr and BLS signatures, whose use is increasing. The signature itself is a 65-byte number, which in combination with a message and a public key can be validated by the signature algorithm.

Double Spend – Is creating two conflicting transactions, one which sends funds to a counterparty, and the other sending those same funds back to yourself. This is prevented by the Bitcoin network and double-spends are not allowed. This is arguably the primary innovation of the Bitcoin blockchain— an algorithm for preventing double-spends. However, in combination with a *51% Attack*, an attacker can cause one conflicting transaction to be replaced with another if he or she controls 51% or more of the hashrate.

DYOR – Stands for "do your own research". This term refers to the need for research and information by an individual or group in the opaque world of cryptocurrencies.

E

Elliptic Curve Cryptography – The preferred public-key cryptography approach for cryptocurrencies to authorize asset transfer. It is favored over older mechanisms based on prime numbers because of the relatively small size of keys and digital signatures and is based on solving equations using an elliptic curve with values in a finite field. The most common elliptic curves used for digital assets are called secp256k1 (e.g., Bitcoin, Ethereum) and ed25519. They are accompanied by an algorithm to create *digital signatures* that can be publicly validated.

ERC-20 – The token standard of Ethereum. ERC-20 tokens are easily exchangeable. Most ICOs use the ERC-20 standard.

Ether – Is a token. Ether tokens are a cryptocurrency created within the Ethereum network. The focus of ether tokens is not as a store of value but rather as a system for creating and paying for the execution of *smart contract* logic

Ethereum – One of the leading cryptocurrencies on the market today after Bitcoin. Ethereum is poised to become larger and more prominent than Bitcoin because of its applicability and adoption by corporate entities. A decentralized, blockchain-based computing platform that allows developers to build and deploy decentralized applications, including smart contracts. In the Ethereum blockchain, mining computers work to earn ether.

Exchange – Cryptocurrency exchanges vary widely in their reliability. An exchange is a platform that allows buyers and sellers to trade a range of digital assets using both fiat currencies and other digital assets. Some exchanges facilitate trading bitcoins for fiat currency, while others enable trading among different digital assets.

F

Fiat Currency – Fiat Fiat money is government-issued currency that is not backed by a physical commodity, such as gold or silver, but rather by the government that issued it. The value of fiat money is derived from the relationship between supply and demand and the stability of the issuing government, rather than the worth of a commodity backing it as is the case for commodity money. Examples would be the US dollar bills, the Japanese yen or the eurozone euro. They are the main alternative currently to cryptocurrencies. The shorthand term is Fiat.

FOMO – Short for "fear of missing out". The fear of missing out creates a bandwagon effect on assets like cryptocurrencies and is one of the major motivating factors for cryptocurrency purchases.

Fork – Occurs when the rules of a blockchain are changed, possibly

creating two (or more) distinct digital assets. This may result from an upgrade to the features of the blockchain, a bug in the consensus algorithm, or changes to the node software. Alternatively, a hard fork may result in a continuation of the network structure if all the participants agree to the changes, install new node software, and update dependent software-like wallets. Soft forks are backward-compatible software updates to a digital asset blockchain. Soft forks do not result in a physical split of the blockchain into two digital assets. An example is when Bitcoin first forked nearly seven years after its introduction in 2017 creating Bitcoin Cash.

FUD – An acronym that stands for "fear, uncertainty, and doubt". This term refers to the forces that can either intentionally or unintentionally bring down the price of an asset. They can be helpful for individuals looking to short an asset.

G

Gas – A fee in Ethereum in ether for every transaction made through the cryptocurrency.

Genesis Block – The first block in a blockchain. As it is the first of the chain, the Genesis Block does not reference any prior block, as all subsequent blocks will.

Going Long – Refers to an investment strategy where investors buys a particular asset like a cryptocurrency with the hopes that the investment will increase in value and that they can sell it later for a profit. This strategy is often employing during a bull market.

Going Short – An investment strategy where an individual borrows a certain amount of an asset to sell and to buy in hopes of paying it back later and lower price. Shorting an asset is a bet that the asset will lose money in the future. This strategy is most effective during a bear market.

Gwei – Another denomination of Ether. Gas prices are usually paid in measurements of Gwei. 1 Ether = 1,000,000,000 Gwei.

H

Halving – Digital asset miners are compensated, or *rewarded,* for their work, which aids the process of validating and processing transactions. In Bitcoin, the reward amount for successfully mining a block is cut in half every four years. This is done to control the distribution of new digital assets in circulation. It is the technical mechanism by which the creator implemented the monetary policy of the system.

Hard Fork – A hard fork is the splitting of a digital asset's blockchain in a backward-incompatible way, resulting in two distinct digital assets. The code and data are replicated from the original digital asset to create the new one, adding backward-incompatible changes. Once the hard fork occurs, the two digital assets are non-fungible with each other but share some transaction and ledger history. Hard forks occur for two key reasons: The first is when competing visions of a digital asset's future development fail to reach agreement. The second is unforeseen bugs or intentional fixes to system-critical issues. When a hard fork occurs, developer and miner support are key components in determining whether the digital assets gain or lose value and relevancy. If poorly implemented, hard forks can also cause instability in the digital asset's network, because of transactions that may be valid on both networks. When used as a feature upgrade mechanism, hard forks require everyone using the digital asset to simultaneously upgrade their node software (called a *flag day*). Coordination of flag days is extremely difficult and, as digital asset networks grow, may become impossible. For this reason, some digital assets such as Bitcoin do not use hard forks as an upgrade mechanism.

Hardware wallet – A secure device for storing cryptocurrency that is particularly effective for cold storage. Hardware wallets are more secure and less flexible than software wallets.

Hash – A hash is the function of mapping data of variable size to a new set of data at a fixed size in such a way that the reverse computation is effectively impossible. Cryptographic hash functions require specific properties to be considered secure, and different digital assets may use different hash functions. The SHA-256 hashing algorithm is used in Bitcoin, and SHA-3 with Ethereum, for example

Hash Rate – When miners run software to create blocks, the algorithm they run is called a *hash*. Miners compute a lot of hashes; the sum of how many hashes they compute in each unit of time is called their *hash rate*. Hash rate is directly correlated with miner earnings.

Increasing one's hash rate by installing new mining devices increases the miner's profits. These computations are special purpose, useful only for mining bitcoin and cannot be repurposed to solve other problems. Hash functions are commonly used for proof-of-work algorithms and are integral to digital signature algorithms. The Bitcoin network must make intensive mathematical operations for security purposes. When the network reached a hash rate of 10 Th/s, it meant it could make 10 trillion calculations per second.

HODL – It started out as a misspelling by a reddit user on a Bitcoin Forum back in 2013 and stuck as a market rally cry from there. It expresses the belief that long-term value is better obtained by holding a digital asset rather than actively trading it.

I

ICO – Short for initial coin offering. An ICO is the process by which a cryptocurrency introduces its methodology and starts the process of building a network and verifying a blockchain.

Institutional Investors – Examples include hedge funds, investment advisors, pensions and endowments, mutual funds, and family offices.

J

Java Cryptography Extension - is an application program interface (API) that provides a uniform framework for the implementation of security features in Java. It was originally developed to supplement the Java 2 Software Developer's Kit (SDK), Standard Edition, versions 1.2.x and 1.3.x, but has since been integrated into the Java 2 SDK, version 1.4.

Japan Virtual Currency Exchange Association – Was Japan's

only recognized self-regulatory organization (SRO) for cryptocurrency assets until April 2020. Its president is Mr. Tateyasu Okuyama. It was renamed in 2020 to the Japan Virtual and Crypto Assets Exchange Association. The Japan Virtual and Crypto Assets Exchange Association (JVCEA) was established in April 2018 in the aftermath of a devastating exchange hack in January 2018, where Japanese exchange lost Coincheck $534 Million NEM tokens. Initially, the JVCEA only assessed the security of crypto-asset exchanges. However, in August that year, the Zaif exchange lost $60 Million in digital assets and led to the JVCEA also issuing restrictive new regulations for "hot wallets", ie. crypto wallets accessible through the internet.

K

Key Pair – The term describes public and private keys used in public-key (or *asymmetric*) cryptography, where the key used to encrypt data is different from the key used to perform decryption. In

Bitcoin, public keys are used as a transaction output in addresses, functioning similarly to an account number or payment instruction, while the private key is known only to the funds' owner and can be used to sign transactions moving those funds. See **Keys**.

Keys – Are long numeric codes that are involved in digital asset transactions, often encoded as hex or alphanumeric strings. *Asymmetric key cryptography* provides a strong security layer in which two different keys are created—a public key that is shared to encrypt a message, and a private key that is confidential to decrypt or sign a message. In Bitcoin these asymmetric keys are used to create digital signatures instead of encryption, which can be validated by everyone. There are two kinds of keys: public and private.

L

Lambo – Short for "Lamborghini". Lambo is a term for what cryptocurrency investors will buy when the investment makes them millionaires.

Ledger – Traditional accounting practices use a ledger to keep track of money movements in and out of accounts, with each party keeping its own ledger and requiring reconciliation between the ledgers of different parties. The Bitcoin network maintains a public ledger that records all transactions. As transactions are executed, updates to the ledger *blocks* containing sets of recent transactions are distributed to millions of computers around the world. Because of the wide distribution of the ledger history, no central point of failure exists, and therefore it is practically impossible for the ledger to be altered by either malice or mistake. The transactions recorded on the Bitcoin ledger are unalterable, permanent, and nearly impossible to erase.

Light Client – A wallet which does not download and validate the full blockchain. Generally, they are wallets (particularly on mobile devices) and rely on a server to supply them with transactions. To have full security for assets, a *full node* is generally required. A light client mechanism was originally proposed by Satoshi Nakamoto called *Simple Payment Verification (SPV)*. Although it was initially deemed to be unworkable, several improvements have been made since. This is an area of active research and development.

Limit Order – An asset arrangement where an individual agrees to buy an asset if its price falls below a certain range and sell an asset if it rises above that range. Limit orders help to minimize risk for investors of any asset including cryptocurrency.

M

Margin Trading – Trading any kind of asset by borrowing money to increase the amount of volume being traded. This approach to trading is one of the riskiest possible.

Market Capitalization / Marketcap – The term *market capitalization* comes from the world of equities and is determined by multiplying the total outstanding shares of an asset by the last available share price. The term has been adopted for use in the digital asset space and is computed by multiplying the total coin supply by the current market value of each coin. Some prefer the term *implied network value*, as the coins are digital assets of decentralized networks rather than shares in a company.

Maximum Coin Supply – This is the total number of coins that can be minted for a particular digital asset. Most digital assets have been designed with caps on the total supply that can be created by the network to drive value by creating digital scarcity. A digital asset's maximum coin supply is a fundamental feature of its design, and some have no fixed maximum supply at all. An example is, Bitcoin's maximum coin supply is set at 21 million.

mBTC – A bitcoin can be split into exceedingly small parts. Each bitcoin is divisible to the eighth decimal place, so each bitcoin can be split into 100,000,000 units (*satoshis*). An mBTC is one thousandth of a bitcoin, or 0.001 BTC. It is also called a *millibitcoin*.

Merkle Tree – Is a binary tree data structure in which a set of data can be compactly committed to so that it cannot be modified. It works by hashing together pairs of data (leaf nodes), hashing the pairs of the pairs from that hashing and so on, in pairs, until there is a single hash remaining. This is known as the *Merkle Root* and is a compact commitment to the entire set of data. Most digital assets use Merkle Trees to ensure that the set of transactions in a block are unmodified. A Merkle Tree also has a feature where by presenting a list of hashes which indicate a branch of the tree, a single element can be proven to be present in the tree. This is the fundamental tool used by Satoshi Nakamoto in his "Simple Payment Verification" (SPV) proposal.

MEW – Stands for MyEtherWallet. A great site that has lots of useful and free tools for Ethereum users, including creating free wallets.

Miner – An individual who mines cryptocurrency. After several years of expansion, miners almost never work on their own and often pool their resources.

Mining – Simply answer, the processing of the blockchain to verify transactions and earn units of cryptocurrency. Mining is the method by which digital assets such as Bitcoin and Ethereum are minted and released into circulation. Mining is also the method by which transactions are incorporated into the blockchain. Finally, mining provides a mechanism to cause the unit of account to acquire a cost of

production, which causes the blockchain to become a financial asset and not just a database entry. Miners perform all the same duties as nodes, and additionally attempt to solve a proof-of-work puzzle that, given a successful solution, gives them the right to publish a block of new transactions and allocate new coins to themselves. They do this by computing a hash repeatedly with different inputs, creating a proof-of-work algorithm. Mining is competitive and requires powerful dedicated hardware, energy consumption, and time.

Mining Pool – Due to the variance of whether a given miner will win a block or not, miners often band together into mining pools. In a mining pool, one node validates transactions and distributes a candidate block to multiple different miners. By agreeing to share winnings if one of the miners in the pool wins the block, pools help reduce variance for its members.

Mining Rig – Any device that an individual uses for cryptocurrency mining. For most individuals, a mining rig is a computer optimized for mining with extra graphics cards. However, mining rigs can be any computer of almost any size or strength.

Mooning – A crypto asset with a price that has gone up massively in a recent period.

N

Node – A computer that possesses a copy of the blockchain and is working to maintain it by connecting and supporting to a cryptocurrency network. It supports it through validation and relaying transactions. At the same time, it also gets a copy of the full blockchain. Any computer that connects to the Bitcoin network, for example, is a node. There are different categories of nodes: **Full nodes** fully enforce all the Bitcoin rules and **Lightweight nodes**, on the other hand, provide for ease of use.

Nonce – A nonce in cryptography is a number used to protect private communications by preventing replay attacks. Nonces are random or pseudo-random numbers that authentication protocols attach to communications. Sometimes these numbers include a timestamp to intensity the fleeting nature of these communications. If subsequent requests to a server, for example during digest access authentication via username and password, contain the wrong nonce and/or timestamp, they are rejected. When used in this way, nonces prevent replay attacks that rely on impersonating prior communications to gain access.

O

Off-Chain Transactions – In blockchain-based cryptocurrencies, off-chain transactions refer to those which occur outside of the blockchain itself and can be contrasted with on-chain transactions. Off-chain transactions can work by swapping private keys to an existing wallet instead of transferring funds, or by using a third-party or coupon-based interlocutor. Off-chain transactions can entail lower fees, immediate settlement, and greater anonymity than on-chain transactions. Depending on the method used, off-chain transactions may eventually have to be recorded on-chain.

P

Peer To Peer (P2P) Network - In a P2P network, the "peers" are computer systems which are connected to each other via the Internet. Files can be shared directly between systems on the network without the need of a central server. In other words, each computer on a P2P network becomes a file server as well as a client. The only requirements for a computer to join a peer-to-peer network are an Internet connection and P2P software. Common P2P software programs include Kazaa, Limewire, BearShare, Morpheus, and Acquisition.

PoS – An acronym that is short for "proof of stake". This will be the future model of Ethereum where owners of ether will be able to make decisions about the currency and have their votes rewarded with more ether.

PoW – An acronym for "proof of work". This approach is the traditional way in which individuals mine cryptocurrency. PoW involves individuals showing their work in mining and verifying the blockchain and being rewarded with cryptocurrency.

Private Key – A private key is a tiny bit of code that is paired with a public key to set off algorithms for text encryption and decryption. It is created as part of public key cryptography during asymmetric-key encryption and used to decrypt and transform a message to a readable format. Public and private keys are paired for secure communication, such as email. Private key is also known as a secret key. In short, sending encrypted messages requires that the sender use the recipient's public key and its own private key for encryption of the digital certificate. Thus, the recipient uses its own private key for message decryption, whereas the sender's public key is used for digital certificate decryption.

Public Key – A public key is created in public key encryption cryptography that uses asymmetric-key encryption algorithms. Public keys are used to convert a message into an unreadable format. Decryption is carried out using a different, but matching, private key. Public and private keys are paired to enable secure communication. A public key may be placed in an open access directory for decryption of the digital signature of the sender, the public key of the message recipient encrypts the sender's message. Public key infrastructure (PKI) produces public and private keys.

Pump and Dump – A scheme where a group of investors place a large amount of money in an asset above its known worth. This "pump" lures in other investors who see the asset gain in value. The original investors then "dump" the stock and its price craters.

Q

QR Code - A quick response (QR) code is a type of barcode that can be read easily by a digital device and which stores information as a series of pixels in a square-shaped grid. QR codes have become more widespread in facilitating digital payments and in cryptocurrency systems such as displaying one's bitcoin address.

Qualitative Analysis – Qualitative analysis uses subjective judgment to analyze a company's value or prospects based on non-quantifiable information, such as management expertise, industry cycles, strength of research and development, and labor relations.

Qualitative analysis uses subjective judgment based on "soft" or non-quantifiable data. Qualitative analysis deals with intangible and inexact information that can be difficult to collect and measure.

R

Ring Signature – Is a type of digital signature that can be performed by any member of a set of users that each have keys. In a peer-to-peer transactions a ring signature enables an individual of a group to sign a transaction without revealing the identity of the actual signer.

S

Satoshi – The smallest fraction of a Bitcoin that can currently be sent. A bitcoin can be divided into one hundred million units. 1 Satoshi = .00000001 Bitcoin

Satoshi Nakamoto – The pseudonym representing a person or group of people who penned the original Bitcoin whitepaper and is the identity credited with inventing Bitcoin itself. Satoshi Nakamoto published a paper in 2008 that jumpstarted the development of cryptocurrency. The paper, *Bitcoin: A Peer-to-Peer Electronic Cash System*,

Scarcity – Most cryptocurrencies include an algorithmically enforced limit on the number of coins or tokens which can be mined. This is different from fiat currency which can be printed at will.

Segregated Witness (SegWit) – SegWit is an action pertaining to Bitcoin that is designed to help increase the block size limit on a blockchain. SegWit helps increase the block size limit by pulling signature data from Bitcoin transactions. Segregate means to separate, and witnesses are the transaction signatures. Hence, segregated witness, in short, means to separate transaction signatures. Segregated Witness was one of many soft-fork upgrades to the Bitcoin network, and it altered the format of transactions.

Selling Pressure – Occurs when the majority of traders or coin holders are selling which indicates that the majority of market believes the market price will decrease.

Sharding – A strategy for scaling upwards where different nonits. des on a cryptocurrency network already have large stretches of the blockchain to work from. Storing strings of blockchain through sharding helps optimize processing power and cuts down on the time required to process new parts of the blockchain.

Shilling – An individual illicitly or blatantly advertising a cryptocurrency. Shilling is seen as more obvious and transparent than normal advertising. It involves someone who over hypes a cryptocurrency that is usually a scam.

Shitcoin – Exactly what it sounds like.

Smart Contract – A contract that governs cryptocurrency transactions and cannot be altered or cancelled. Smart contracts often have conditions and stipulations attached to different transactions.

Soft Fork – In the world of cryptocurrencies, a soft fork is where the cryptocurrency chain experiences a divergence. A soft fork is a change made to cryptocurrency technology creating a temporary split in the group of recordings (blockchain). This change creates all new, valid recordings (blocks) that are slightly different from the original blocks. They are simply different enough that users of the new technology see blocks from original technology as invalid. But, users of the original technology see no problem with either one. Soft forks can refine the governance rules and functions of a digital asset blockchain but, unlike hard forks, are compatible with the previous blockchain. For a soft fork to be implemented, a specific level of readiness to enforce the new rules must be signaled by miners.

Software Wallet or Soft Wallet – A program that stores certain public and private keys used to buy, sell, and hold cryptocurrencies. The software wallet is the application that ensures cryptocurrency users can transfer and make purchases with his or her cryptocurrency.

Stable Coin – Stable coin is a term for trusted cryptocurrencies that are not small, untrusted upstarts like altcoins.

Store-Of-Value (SOV) – Monetary economics is the branch of economics which analyses the functions of money. Storage of value is one of the three generally accepted functions of money. The other functions are the medium of exchange, which is used as an intermediary to avoid the inconveniences of the coincidence of wants, and the unit of account, which allows the value of various goods, services, assets, and liabilities to be rendered in multiples of the same unit. Money is well-suited to storing value because of its purchasing power. It is also useful because of its durability. An asset is a "good "store of value" if the purchasing power does not degrade over time.

Support Level – This refers to the price point in which downward price movement is resisted due to market conditions.

T

The Flippening – The moment when Ethereum becomes larger than Bitcoin. There are websites that currently count down to this moment.

Token – A representation of an asset or utility that resides on top of another blockchain (like Ethereum). Tokens are different from coins in that coins typically operate on their own blockchain (like Bitcoin or Ethereum).

Total Circulating Coin Supply – This is the total number of coins that a particular digital asset has in circulation.

Total Coin Supply – This is the total number of coins that have been minted for a particular digital asset, although not all coins minted may be in circulation.

Trading Volume – Is the total amount of digital coins that were traded during a certain period.

Transaction Fee – A transaction fee is an amount of cryptocurrency that is attached to a transaction and that incentivizes miners to process the user's transaction. Users can choose how much to pay for their

transactions to be processed. That is why during times of network congestion, the average transaction fee goes up, as users are trying to incentivize miners to process their transactions over other users' transactions.

Two Factor Authentication or 2FA – A second layer of identity verification to secure your account when logging in. It usually requires you to enter a unique code sent to your mobile phone during log in. This prevents hackers from accessing your account with a stolen username and password since they would need your phone to authenticate the log in.

U

uBTC – A bitcoin can be split into very small parts. A uBTC is one millionth of a bitcoin, or 0.000001 BTC. It is also called a *microbitcoin*.

Unspent Transaction Output – A UTXO is the amount of digital currency remaining after a cryptocurrency transaction is executed. UTXOs are processed continuously and are responsible for beginning and ending each transaction. When a transaction is completed, any unspent outputs are deposited back into a database as inputs which can be used later for a new transaction.

V

Value at Risk (VaR) – Is a statistic that measures and quantifies the level of financial risk within a firm, portfolio, or position over a specific time frame. This metric is most used by investment and commercial banks to determine the extent and occurrence ratio of potential losses in their institutional portfolios. Risk managers use VaR to measure and control the level of risk exposure. One can apply VaR calculations to specific positions or whole portfolios or to measure firm-wide risk exposure.

W

Wallet – A digital asset wallet, or "E-Wallet" is a piece of software based system that maintains keys and manages addresses. It also stores users payment information and passwords for numerous payment methods and websites. A wallet is comprised of a set of addresses. If the wallet has the private keys for these addresses, it can send transactions. If it does not have the private keys for these addresses, it is called a *watch-only wallet*, as might be used by an auditor.

Wei – The smallest denomination of Ether. 1 Ether = 1,000,000,000,000,000,000 Wei.

Whale – An individual who owns a large amount of an asset such as a cryptocurrency. Whales have a disproportionate impact on the price and volatility of an asset.

X

XBT – BTC is the original shorthand for *bitcoin*, there has been an increase in the use of the term XBT. The reason for this is that the International Organization for Standardization (ISO), which keeps a listing of all currencies, uses *X* to symbolize a currency that is not attached to a specific country.

Z

Zero Knowledge Proofs – Are an experimental technology that allows one to cryptographically prove a statement, without revealing the input data. ZKPs are being actively explored by several blockchain and cryptocurrency projects and are a fundamental piece of engineering infrastructure in the space.

MISC.

51% Attack – In the Bitcoin Whitepaper, Satoshi Nakamoto computed the probability that transactions could be reversed. The ability to reverse transactions is only possible probabilistically if no entity has more than 51% of the mining hash rate and supports the rule of thumb to wait 6 confirmations before considering a transaction settled, as well as the concept of a 51% attack. If any entity controls 51% or more of the hashrate, they can arbitrarily censor transactions and/or prevent progress, though they cannot directly steal funds. Theft of funds is possible by such an entity only if a counterparty follows the 6 confirmation rule, the attacker has 51% of the hash rate, and the attacker creates a double spend.

Digital Money, A New Frontier

The concepts of transferring value have been evolving ever since primitive cultures accepted cowry shells in Africa, large stone wheels on the Pacific island of Yap, and strings of beads called wampum used by Native Americans as a means of monetary exchange. While the medium of exchange has evolved, the fundamental principles of money have remained consistent. The six characteristics of money are still:

1. Durability
2. Portability
3. Divisibility
4. Uniformity
5. Limited Supply
6. Acceptability

Even though there have been many forms of money in history, some forms have been more useful than others because they applied these characteristics more effectively. While there is a list of six characteristics to money, there is also a list of the six functions of money:

1. Medium of Exchange – Replaces the barter system.
2. Measure of Value - By acting as a common denominator it permits everything to be priced, that is, valued in terms of money. Thus, people are enabled to compare different prices and thus see the relative values of different goods and services.
3. Store of Value - As a store of purchasing power. It can be held over a period of time and used to finance future payments.

4. The Basis of Credit - Facilitates loans for borrowers use to obtain goods and services when they are needed most.

5. A Unit of Account - The implication is that money is used to measure and record financial transactions as also the value of goods or services produced in a country over time. The money value of goods and services produced in an economy in an accounting year is called gross national product.

6. A Standard of Postponed Payment - Again used as a medium of exchange, but this time the payment is spread over a period of time. its development.

To better understand the overall Cryptocurrency market an investor needs to understand these basic principles of money because Bitcoin was initially created to be Digital Money. Even though the technology of Bitcoin is considered old and clunky by today's technical standards it has established itself as a "Store of Value" and a "Medium of Exchange." While it might not be the medium of exchange for the retail investor, might certainly be for the global financial institutions, corporations, and governments. When it comes to a medium of exchange for the retail investor and the common consumer the Altcoin segment of the market will more than likely lead the way. As one begins to investigate this new digital frontier it becomes more clear that the emergence of digital assets signal a fundamental change in the way value will be transferred globally in the near future.

History of Digital Cash

While many of us might think that digital money made its debut around 2009 with Bitcoin, the truth is that the concepts of digital money go back a couple of decades to 1989. In actuality, the world of electronic encryption goes back as far as 1952 with the **TSEC/KL-7** which was an off-line non-reciprocal rotor encryption machine used by the US

Military up to mid- 1960's. The KL-7 had rotors to encrypt the text, most of which moved in a complex pattern, controlled by notched rings. As technology improved, new machines were introduced like the **TSEC/KW-26** and the **TSEC/KG-84** by the early 1908's. While these are "fun facts" about military security, the point is that electronic encryption has been around for over a half century. While it may have started as a military function it was just a matter of time before the technology moved into the commercial markets.

DigiCash was one of the earliest electronic money companies. Clearly a pioneer in its time, it was a company founded by an electric currency innovator named David Chaum. He earned his doctorate in computer science from the University of California, Berkeley in 1982. David Chaum's dissertation "Computer Systems Established, Maintained and Mutually Trusted by Suspicious Groups" is considered by many as a prototype of blockchain technology. Later in 1982 he published a revolutionary paper on the technology of anonymous cash transfers called, "Blind Signatures for Untraceable Payments."

DigiCash was in business from 1989 to 1998 where Chaum developed several cryptographic protocols which administered DigiCash transactions and set his currency apart from its competitors. Often referred to "Chaumian eCash," these protocols made DigiCash an important predecessor of modern digital currencies. Unfortunately, he was unable to convince banks to adopt its technology and filed for bankruptcy in 1998. Keep in mind that was only ten years before the financial crisis of 2008 which was one of the catalysts for the development of blockchain-based cryptocurrencies like Bitcoin.

HashCash, in 1997, was birthed by a renowned crypto expert and computer hacker named Adam Back was working on inventions of his own. The US Postal Service was finding itself in competition with a new digital protocol called "Emailing." The new technology brought with a problem called "spamming" which Adam Back wanted to fix, hence the origins of HashCash. Back developed the "proof of work" system. This as a countermeasure against spam in emails and in blogs. (spam is spam). Back devised this technology based on a job called; "Assessment through processing or combating spam" a document which was published in 1992.

While HashCaah is not really a digital cash its greatest use in computing, the mechanism to control spam attacks on mail services, has become part of the foundation of the cryptocurrency technology. It offers two use cases. The first, is the protection of the connection between the client and the server. The second, is in utilization within file systems where hashing functions create a unique signature for each block of stored data which enables the system to verify their authenticity. Bitcoin uses the HashCash's Proof-of-Work function as the mining core. All bitcoin miners whether CPU, GPU, FPGA or ASICs are expending their effort creating HashCash proofs-of-work which act as a vote in the blockchain evolution and validate the

```
ith_grid(name, label_p, label_n, oracle, n_features, depth,
--------- retrain in X with grid ................
nge(50, 601, 50):
nlineBase(name, label_p, label_n, oracle, n_features, depth,
lect_pts(QSV, -1)
rnelRetraining(name,
                online.get_QSV(), online.get_QPT selected,
                online.get_QSV(), online.get_QPT selected,
                test_x, test_y,  # test data
                n_features)

SV=%d, Q=%d, ' % (QSV, online.get_n_query()), end=print output

with_grid(name, label_p, label_n, oracle, n_features, depth
--------- retrain in F with grid ---------------
```

blockchain transaction log. Without this software, the current Bitcoin mining and security structure would not be possible.

B-Money was first unveiled in 1998 essay by computer scientist Wei Dai. Dai was a computer engineer and graduate from the University of Washington. He intended B-Money to be an anonymous, distributed electronic cash system which was conceived to have several of the same services and features we find in blockchain technology today.

Dai's B-Money concepts included the requirement for computational work in order to facilitate the digital currency, the stipulation that this work must be verified by the community in a collective ledger, and rewarding workers for their input. To ensure that transactions remained organized, Dai proposed that collective bookkeeping would be necessary, with cryptographic protocols helping to authenticate transactions which led to suggesting the use of digital signatures, or public keys, for authentication of transactions and enforcement of contracts. B-Money was never officially launched, but the work of Wei Dai has become very influential in the technical development of cryptocurrency. Some would strongly suggest that is a part of the foundation for Bitcoin.

Bit Gold was a proposal introduced in 1998 and is recognized as one of the earliest attempts at creating a decentralized virtual currency. Its author was another blockchain pioneer Nick Szabo. Mr. Szabo is a computer scientist and well-known cryptographer because of his work and research in digital contracts and currency. He is a computer science graduate from the University of Washington in 1989, law graduate from George Washington University Law School and an honorary professor at Universidad Francisco Marroquin. Szabo's goal was to develop the concept of "smart contracts" by bringing what he called "highly evolved" practices of contract law and practice to the design of electronic commerce protocols between strangers on the internet. Smart Contracts are a major technical feature in cryptocurrency development.

The similarities between Bit Gold and Bitcoin and Bit Gold have led to speculation is Nick Szabo, Satoshi Nakamoto. Here are a few couple examples: First, the Bit Gold structure, requires users to solve a cryptographic puzzle using computing power. All solved puzzles are sent through a Byzantine Fault Tolerant (BFT) peer-to-peer network and then assigned to the public key of the puzzle solver. Secondly, Every solution then becomes a part of the next puzzle, creating a chain that links the most recent puzzle's solution to the outcome of the following one, thereby validating blocks of transactions. Lastly, Bit Gold combines different elements of cryptography and mining to accomplish decentralization, including time-stamped blocks that are stored in a title registry and are generated using proof-of-work (PoW) strings. Even though this project was never implemented it is widely considered to be the precursor to Bitcoin's protocol.

Reusable Proof-of-Work (RPOW) was a 2004 invention by Hal Finney II with the intention for it to be a prototype for a digital cash. Harold Finney II graduated from California Institute of Technology with a BS in engineering in 1979. He immediately went to work in the computer gaming field and ended his career working for PGP Corporation a computer software company. He was also a noted cryptographic activist. Although never intended to be more than a prototype, RPOW was a very sophisticated piece of software that would have been capable of serving a huge network and was a significant early step in the history of digital cash and Bitcoin.

So how does this software work? An RPOW client creates an RPOW token by providing a proof-of-work string of a given difficulty, signed by his private key. The server then registers that token as belonging to the signing key. The client can then give the token to another key by signing a transfer order to a public key. The server then duly registers the token as belonging to the corresponding private key. In the use of digital cash the "double spending" problem is fundamental. RPOW solves this problem by keeping the ownership of tokens registered on a trusted server. However, RPOW was built with a sophisticated security model intended to make the server managing the registration of all RPOW tokens more trustworthy than an ordinary bank. This was a noteworthy advance in the 25yr search for viable decentralized money, but still did not solve the issue of fairly creating and distributing the initial supply of coins, that would come later in the development of Bitcoin.

In 2009, Hal Finney II was the first recipient of a Bitcoin Network transaction. He had some utopian ideas about the use of computers which can be found on old Cypherpunk feeds. It was obvious to Finney, that in the face of problems relating to loss of privacy, creeping computerization, massive databases, more centralization David Chaum offered a completely different direction to go in, one which puts power into the hands of individuals rather than governments and corporations.

Finney honestly believed the computer can be used as a tool to liberate and protect people, rather than to control them. It is ideas like these which have helped shape this new frontier of digital money.

For me this has been brief but interesting history lesson on digital cash. I can only imagine what I would find if a did a deep dive in my research. My immediate thoughts are, first the admiration for the intelligent minds behind the technology and secondly, I am excited to see what new technological developments come up in the future. These early projects obviously had limitations and challenges. However, they become the steppingstones for other great minds to build upon and which took the technology to the next level of cryptography. Each one of these projects, and others not listed, made contributions which led to breakthroughs in decentralized networking and   protocols enabling this new frontier of digital money to become reality.   Like every new frontier story there will be calamities and good fortune. As characters in this narrative, we get to decide which side of the digital coin we want to be on.

Digital Money & Bitcoin Today

Back to the topic of money. Most of us think of money as the conventional banknotes, or fiat currency, we use every day as a medium of exchange. However, as technology has advanced, we have slowly moved into a new realization that our "hard earned cash" has become is a series of digits on a digital record which is controlled by large financial institutions. Payments today are mostly facilitated by financial intermediaries like banks, credit unions and other financial service companies who we trust are making accurate and secure changes to our records. As our financial records have become digital new mediums of exchanging value have emerged.

Today there are various of payment methods and stores of value to

choose from which are exclusively digital in nature. Companies like Veritran, Paytm, Apple Pay Cash, Paypal, Square Cash and Zelle just to name a few enable mobile payments for millions of merchants and consumers to exchange value for goods and services globally with the press of a few buttons. With all these applications at our fingertips we still need centralized financial institutions to agree to the transaction records among all parties to complete the transfer of value. Essentially, these applications serve as an innovative user interface which has been built upon traditional financial payment systems which are antiquated and work by translating manual record keeping protocols and turning them into an electronic format.

Bitcoin and other digital assets have come to open the door to a new paradigm of how we think about and use money. In this utopian world of finance, digital assets are poised to become new investment assets and stores of value which will be tradable on licensed global exchanges and accessible to both institutions, corporations, and individuals equally. This new financial global frontier is a decentralized model whose framework relies on and is powered by distributed ledgers which eliminate the need for centralized financial institutions to facilitate the exchange of value. As Bitcoin and the other Altcoin projects emerge and become adopted into mainstream use the question arises, why do we need fiat currency?

Bitcoin and other altcoins with fixed supply schedules and monetary policies begin to illuminate the truth that fiat currencies have no intrinsic value, but rather nothing more than artistic pieces of sophisticated paper. The traditional global monetary system is going through a transition because the old protocols have reached a culmination point. New technologies are introducing new protocols, the US Dollar as the global reserve currency is being challenged, and another wealth transfer is taking place before are very eyes if we are paying attention to the signals and not the latest trends in pop culture. This new class of digital currencies is offering us an opportunity to live out some of the utopian notions Hal Finney was trying to awaken us to back in 2004.

Digital Money's Evolution

As developers, computer scientist and cryptographers flock to the new frontier the ecosystem will grow, new opportunities will emerge and innovative applications for digital assets will appear. As companies like IOHK and the Cardano Foundation partner with educational institutions like the University of Wyoming pioneering concepts will flourish bringing new excitement to the whole ecospace not just in the finance industry. MIT's Digital Currency Initiative mission is to create a future in which moving value across the internet is as intuitive and efficient as moving information. They raise a salient point that blockchain technology is being compared to the Internet. The Internet

enabled people to easily call each other without a phone company, send documents without a mail carrier, or publish an article without a newspaper.

Blockchain technology is a decentralized public ledger of debits and credits built on the principle that no one person or company owns or controls it. The intent is that users control their data directly. This new system promises to let people transfer money without a bank, write simple, enforceable contracts without a lawyer, or perhaps even turn personal property such as real estate into digital assets which can be transferred or pledged with near-zero transaction fees.

The Internet was officially birthed on January 1, 1983. 38 years later, there are 4.72 billion internet users in the world today. Globally, internet user numbers are growing at an annual rate of more than 7.5 percent. Bitcoin is only 12 years old. Just imagine what could happen in this new Digital Money frontier in the next two and half decades. Like in any new frontier, there will be villains and heroes. We all will need to pick a side proactively or by default, the choice is ours.

For myself I openly admit that understanding, and at times using, technology does not come easy for me. At first, through ignorance and fear I rejected this new frontier only to discover the foolishness of my decision. However, I have decided to admit my shortcomings and embrace my fears with education and diligence. I have chosen to associate with the innovators and to the best of my ability become a champion for blockchain technology. As an investor, I believe this new posture towards technology will yield benefits and blessings for myself and my children in years to come. What choice will you make? Will you join me in the quest of discovering this new frontier of Digital Money.

Newbies become Masters

Novicius – fit – Dominus

AUTHORS BIO

Christopher Brown - started his writing career back in 2006 when he began contributing articles to the *Long Beach Journal, L.A. Times* and *Variety.* This side hobby evolved into becoming a regular contributor for an online publication called *The Power Player Lifestyle* magazine. In 2012, he published his first book entitled *Financial Crisis or Faith?* and his second in 2020 called *MoneySeeds.*

The first book inspired him to create a Christian personal finance non-profit called FaithFinance. As the Executive Director, Christopher is committed to building a community of believers who desire to live a lifestyle of stewardship. As a certified personal Financial Coach, he is driven to encouraging, educating and empowering others to "Live Faithfully, so that they can Live Abundantly." In keeping with the mission of FaithFinance, CryptoNovus was birthed as a specialty area of personal finance. As the world of digital money evolves people will need quality education and a resource founded on integrity in order to navigate this new frontier. FaithFinance will be that resource.

Christopher Brown is on the board of directors for a non-profit Christian ministry called Restored Hope, Inc and is a member of Cottonwood Church in Los Alamitos, California. Professionally, he is a licensed architect, interior designer and construction project manager in California with a bachelor's degree in architecture from California Polytechnic University, San Luis Obispo. Being recently remarried, and a father of three, Christopher strongly believes his greatest accomplishment is having those closest to him know they are loved by him. As a native Southern Californian, Christopher loves the warm sun and the beautiful California coastline ... it is where he replenishes his soul.